John S. Siebert, Frederic C. Biggin

Modern Stone-Cutting and Masonry

With special reference to the making of working drawings. First Edition

John S. Siebert, Frederic C. Biggin

Modern Stone-Cutting and Masonry
With special reference to the making of working drawings. First Edition

ISBN/EAN: 9783337882396

Printed in Europe, USA, Canada, Australia, Japan

Cover: Foto ©Paul-Georg Meister /pixelio.de

More available books at **www.hansebooks.com**

MODERN

STONE-CUTTING AND MASONRY.

*WITH SPECIAL REFERENCE TO THE MAKING
OF WORKING DRAWINGS.*

BY

JOHN S. SIEBERT, C.E.,

AND

FREDERIC CHILD BIGGIN, B.S.,

Instructor in Architecture, Lehigh University.

FIRST EDITION.

FIRST THOUSAND.

NEW YORK:

JOHN WILEY & SONS.

LONDON: CHAPMAN & HALL, LIMITED.

1896.

PREFACE.

— —

IT has long been recognized that the works on stone-cutting used in this country are entirely out of date and not abreast of the present demands.

The reasons are apparent. Those works were modelled strictly on lines set by the French school, which dealt chiefly with the heavy masonry of fortifications, and the intricate stone constructions used for the immense churches and cathedrals, built when the Church throughout Europe was in the zenith of its power. Times have changed. The heaviest masonry is no protection against modern projectiles, and steel frames veneered with stone and brick form the walls of our largest buildings. This is not the place to discuss the merits of the different styles of construction, the grace of the stone arch and the steel cantilever, or to place the beauty of a stone carving in opposition to the hard lines of iron ornament; suffice it to say that first cost is a very important factor in this country, and that, stability and usefulness being equal, the cheaper structure is generally adopted.

Intricate stonework is therefore excluded everywhere, save on very costly structures, and a treatise dealing with such work would be interesting and valuable only in so far as it presented problems to be worked by the student in search of exercise for his mental powers.

It has been the aim of the authors, in this little volume, to collect and make available for the student's use only such material as finds a direct application in engineering and architectural practice in this country. Chapter I describes the mason's and stone-cutter's tools, the shape and finish produced by these, the classification of masonry according to shape and finish, as defined by the best existing authorities, and closes with a few general remarks on bond, dressing, etc.

Chapter II deals with stone-cutting and masonry proper, and after a few preliminary definitions and descriptions takes up the study of the different examples chosen. The explanations are detailed where thought necessary, but some of the drawings are intended only as models according to which the student should make similar ones.

The skew or oblique arch is not treated here, because it is rarely used; the right arch with skew face, Plate VIII, or some other simple expedient being substituted. If it becomes necessary to build a true skew arch, the student should refer to "Traité Pratique de la Coupe des Pierres," by Émile Lejeune, Paris; the German works of Dr. Ringleb; and the American publications of Prof. Warren bearing on the subject.

The drawings should in all cases be made at least to a scale of one fourth inch per foot, with details three times this size; better still if both can be made double the scales mentioned, although it should be borne in mind that those first given are the ones commonly used in practice. Let the work be executed accurately, neatly, and with strong lines, remembering that nearly all office drawings are nowadays made on tracing-linen, for blue-printing, and that fancy titles are not in demand. The draughtsman who can make the plain Roman letters well has a broad foundation for all styles of lettering.

For exercises other than those given in the book the teacher should select some good existing masonry structures, have the class take full measurements of them, and then make complete drawings in accordance with the samples given in these pages. Models should always be available for the student, and the authors have in preparation a series of models illustrating various stone-cutting problems. They may be obtained direct from the undersigned.

The authors desire to acknowledge their indebtedness to Mr. Bernard R. Green, superintendent and engineer of the building for the Library of Congress; Messrs. Webster and Shaw, engineers of the Lehigh Valley Railroad; and Messrs. E. S. Wheeler and J. L. Callard, United States engineers on the Sault Ste. Marie Canal, for the valuable material furnished for use in this book.

JOHN S. SIEBERT.
FREDERIC CHILD BIGGIN.

SOUTH BETHLEHEM, PA.,
Jan. 1, 1896.

TABLE OF CONTENTS.

CHAPTER I.

DEFINITIONS AND CLASSIFICATIONS.

CHAPTER II.

STONE CUTTING AND MASONRY.

MODERN STONE-CUTTING AND MASONRY.

CHAPTER I.

DEFINITIONS AND CLASSIFICATIONS.

ART. I.—NAMES AND DESCRIPTIONS OF STONE-MASON'S TOOLS.

1. The following definitions of stone-mason's tools and masonry are taken from the report of the Committee of the American Society of Civil Engineers, prepared to secure uniformity of nomenclature on this subject. (See Trans. Am. Soc. C. E., vol. v. I. pp. 297–304.)

2. The tools used at the quarry for roughly shaping the stones for transportation and also for shaping for the lower grades of masonry are (see Fig. 1): the *Double-face Hammer*, a tool weighing from 20 to 30 pounds; the *Face-hammer*, having one blunt and one cutting edge and weighing less than the double-face hammer; the *Cavil*, having one blunt and one pointed or pyramidal end and weighing from 15 to 20 pounds; the *Pick* somewhat resembles the pick used in digging, is mostly used on limestone and sandstone, and is from 15 to 24 inches long, the thickness at the eye being about 2 inches.

The tools shown in Figs. 2–6 are used for dressing proper.

The *Axe* or *Pean-hammer* (Fig. 2) has two opposite cutting edges. It is used for making drafts around the arris or edge of stones, and in reducing faces and sometimes joints to a level. Its length is about 10 inches and the cutting

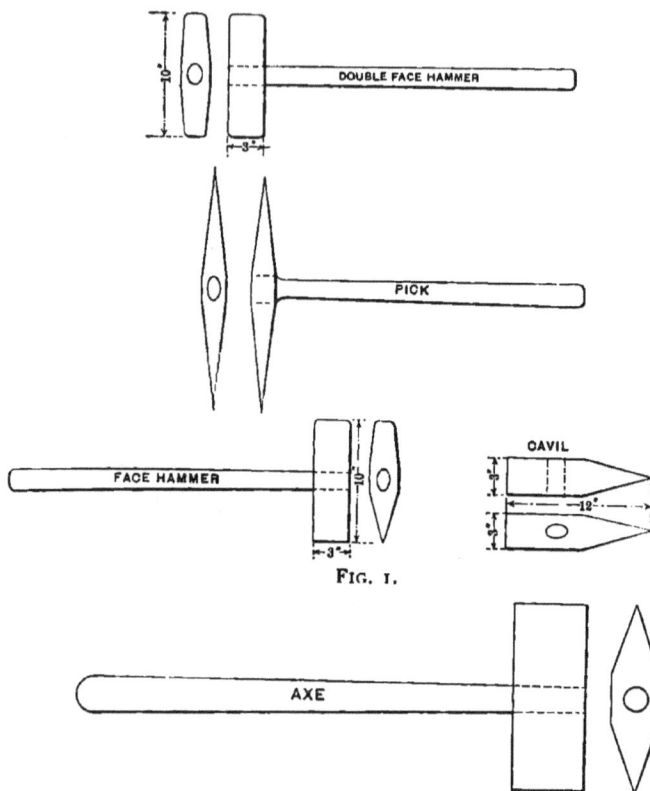

FIG. 1.

FIG. 2.

edge about 4 inches. It is used after the point and before the patent hammer.

The *Tooth-axe* (Fig. 3) is like the axe, except that its cutting edges are divided into teeth, the number of which vary with the kind of work required. This tool is not used in granite and gneiss cutting.

The *Bush-hammer* (Fig. 4) is a square prism of steel whose ends are cut into a number of pyramidal points. The length of the hammer is from 4 to 8 inches, and the cutting face from 2 to 4 inches square. The points vary in number

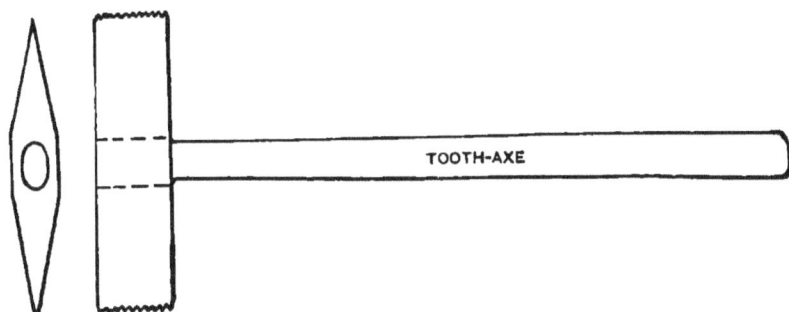

TOOTH-AXE

Fig. 3.

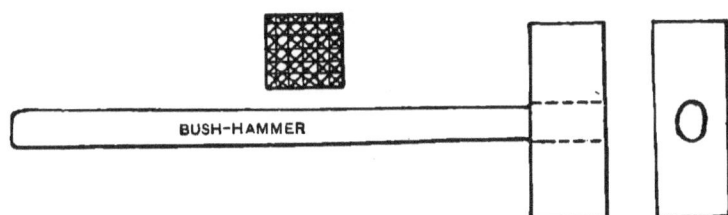

BUSH-HAMMER

Fig. 4.

and in size with the work to be done. One end is sometimes made with a cutting edge like that of an axe.

The *Crandall* (Fig. 5) is a malleable-iron bar about 2 feet

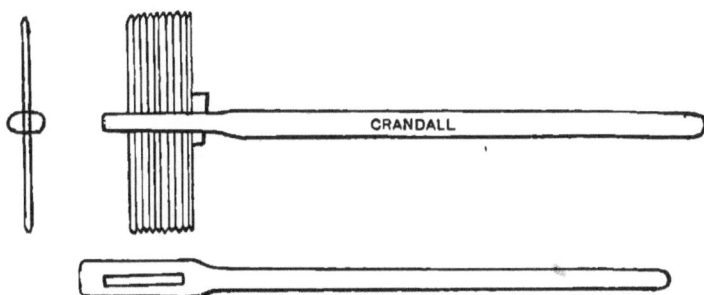

CRANDALL

Fig. 5.

long, slightly flattened at one end. In this end is a slot, 3 inches long and ⅜ inch wide. Through this slot are passed ten double-headed points of $\frac{1}{4}$-inch squared steel, 9 inches long, which are held in place by a key.

The *Patent Hammer* (Fig. 6) is a double-headed tool so formed as to hold at each end a set of wide thin chisels.

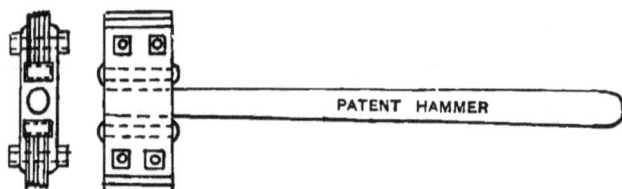

PATENT HAMMER

Fig. 6.

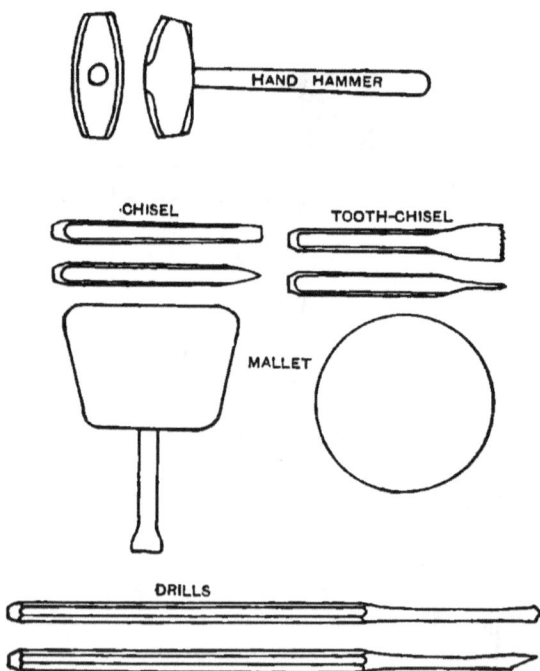

HAND HAMMER

CHISEL   TOOTH-CHISEL

MALLET

DRILLS

Fig. 7.

The tool is in two parts, which are held together by the bolts which hold the chisels. This tool is used for giving a finish to the surface of stones.

3. All of the above mentioned are two-handed, or require

both hands of the workman to use them. The remaining tools to be described require the use of only one hand for each.

The hand hammer and mallet are used for chiselling hard and soft stones respectively. The pitching-chisel is used to make a well-defined edge on the face of a stone, and the point for finishing a surface. The chisel and tooth-chisel are used for cutting drafts or margins on the face of stones, but the latter is used only on marbles and sandstones. The splitting-chisel and plug are used for splitting stratified and unstratified stone respectively.

In architectural carving a variety of chisels of different forms are used, for most of which no specific name exists.

For extensive operations machinery replaces many of the hand tools above described. The chief ones are the saw and

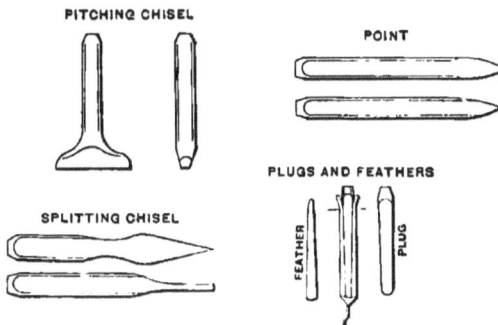

FIG. 8.

planer, the latter being a machine very similar to a metal planer. By simply changing the cutting tool a great variety of surfaces may be formed. A certain amount of hand-work, however, always remains to be done on nearly all stones.

ART. 2.—STONES CLASSIFIED ACCORDING TO FINISH.

4. All stones used in building come under one of three classes, viz. :

I. Rough stones that are used as they come from the quarry.

II. Stones roughly squared and dressed.

III. Stones accurately squared and finely dressed.

In practice the line of separation between them is not very distinctly marked, but one class gradually merges into the next.

5. I. Unsquared Stones or Rubble.—This class covers all stones which are used as they come from the quarry without other preparation than the removal of very acute angles, and excessive projections from the general figure. The term " backing," which is frequently applied to this class of stone, is inappropriate, as it properly designates material used in a certain relative position in a wall, whereas stones of this kind may be used in any position.

6. II. Squared Stones.—This class covers all stones that are roughly squared and roughly dressed on beds and joints. The dressing is usually done with the face hammer or the axe, or in soft stones with the tooth-axe. The distinction between this class and the third lies in the degree of closeness of the joints which is demanded. Where the dressing on the joints is such that the distance between the general planes of the surfaces of adjoining stones is one half inch or more, the stones properly belong to this class.

Three subdivisions of this class may be made, depending on the character of the face of the stone:

(*a*) *Quarry-faced* stones are those whose faces are left untouched as they come from the quarry.

(*b*) *Pitch-faced* stones are those on which the arris is clearly defined by a line beyond which the rock is cut away by the pitching-chisel, so as to give edges that are approximately true. (Fig. 20.)

(*c*) *Drafted* stones are those on which the face is surrounded by a chisel-draft, the space inside the draft being left rough. Ordinarily, however, this is done only on stones in which the cutting of the joints is such as to exclude them from this class.

In ordering stones of this class the specification should always state the width of the bed and end joints which are expected, and how far the surface of the face may project beyond the plane of the edge. In practice the projection varies between 1 inch and 6 inches. It should also be specified whether or not the faces are to be drafted.

7. **II. Cut Stones.**—This class covers all squared stones with smoothly dressed beds and joints. As a rule all the edges of cut stones are drafted, and between the drafts the stone is smoothly dressed. The face, however, is often left rough when the constructions are massive.

Rough-pointed.—When it is necessary to remove an inch or more from the face of a stone, it is done by the pick or heavy points until the projections vary from $\frac{1}{2}$ to 1 inch. The stone is then said to be rough-pointed. This operation precedes all others in dressing limestone and granite.

Fine-pointed.—If a smoother finish is desired, rough-pointing is followed by fine-pointing, which is done with a fine point. It is only used where the finish made by it is to be final, and never as a preparation for final finish by another tool.

Crandalled.—This is only a speedy method of pointing, the effect being the same as fine-pointing, except that the dots are more regular. The variations of level are about $\frac{1}{8}$ inch, and the rows are made parallel. When other rows at right angles to the first are introduced, the stone is said to be cross-crandalled.

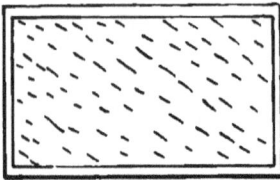

FIG. 9.

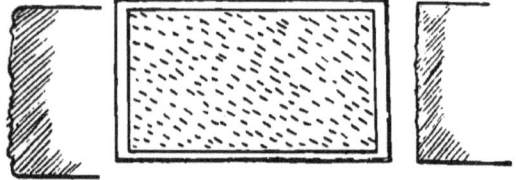

FIG. 10.

Axed or Pean-hammered, and Patent-hammered.—These two vary only in the degree of smoothness of the surface

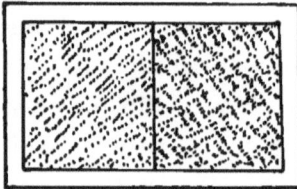

FIG. 11.

FIG. 12.

which is produced. The number of blades in a patent hammer varies from 6 to 12 to the inch, and in precise specifications the number of cuts to the inch must be stated, such as 6 cut, 8 cut, 10 cut, 12 cut. The effect of axing is to cover the suface with chisel-marks which are made parallel as far as possible. Axing is a final finish.

Tooth-axed.—The tooth-axe is practically a number of points, and it leaves the surface of the stone in the same condition as fine-pointing. It is usually, however, only a preparation for bush-hammering, and the work is then done without regard for effect so long as the surface of the stone is sufficiently levelled.

Bush-hammered.—The roughnesses of a stone are pounded

off by the bush-hammer, and the stone is then said to be
"bushed." This kind of finish is dangerous on sandstone,
as experience has proved that sandstone thus treated is very
apt to scale.

In dressing limestone which is to have a bush-hammered

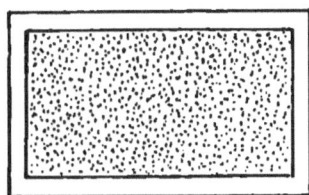

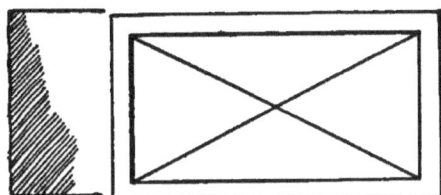

FIG. 13.   FIG. 14.

finish the usual sequence of operations is: 1st, rough-pointing; 2d, tooth-axing; 3d, bush-hammering.

Rubbed.—In dressing sandstone and marble it is very
common to give the stone a plane surface at once by the use
of the stone-saw. Any roughnesses left by the saw are re-
moved by rubbing with grit or sandstone. Such stones
therefore have no margins. They are frequently used in
architecture for string-courses, lintels, door-jambs, etc., and
they are also well adapted for use in facing the walls of lock-
chambers, and in other localities where a stone surface is
liable to be rubbed by vessels or other moving bodies.

Diamond Panels.—Sometimes the space between the margins is sunk immediately adjoining them and then rises
gradually until the four planes form an apex at the middle of
the panel. Such panels are called diamond panels, and in
the case described the panel is a sunk diamond panel. When
the surface of the stone rises gradually from the inner lines of
the margins to the middle of the panel, it is called a raised
diamond panel. Both kinds of finish are common on bridge

quoins and similar work. The details of this method of
dressing should be given in the specifications.

8. The student should take care not to confound the
above-described three classes with what is termed first-,
second-, and third-class masonry in railroad-engineering work.
Extracts from specifications defining these classes are given
in §§ 45, 53, 54.

ART. 3.—DEFINITIONS OF PARTS OF THE STRUCTURE.

9. *Face*, the front surface of a wall; *back*, the inside
surface.

Facing, the stone which forms the face or outside of the
wall.   *Backing*, the stone which forms the back of the wall.
Filling, the interior of the wall.

Batter.—The slope of the surface of the wall.   A batter
of 1 inch to 1 foot, etc., means a horizontal departure of one
inch from a vertical direction for each foot of altitude.

Course.—A horizontal layer of stone in the wall.   If the
stones of each layer are of equal thickness throughout, it is
termed regular coursing; if the thicknesses are unequal, the
term random or unequal coursing is applied.

Joints.—The mortar layer between the stones.   The
horizontal joints are called bed-joints or simply beds; the
vertical joints are sometimes called the builds.   Usually the
horizontal joints are called beds, and the vertical ones joints.

Coping.—A projecting course of heavy stones (generally
weathered) on the top of the wall to protect it.   (A coping is
weathered when it is bevelled sufficiently to throw the water
away from the face of the wall.)

Pointing.—A better quality of mortar put in the face of
the joints to help them resist the influence of the weather.

10. *Bond.*—The arrangement of stones in adjacent courses. (See Art. 5.)

Stretcher.—A stone whose greatest dimension lies parallel to the face of the wall.

Header.—A stone whose greatest dimension lies perpendicular to the face of the wall.

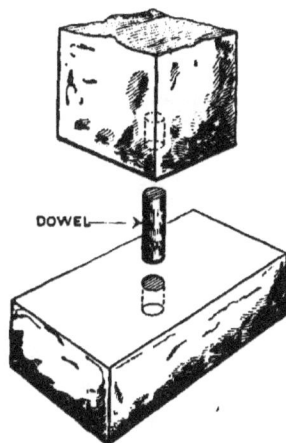

FIG. 15.

Dowels.—Pins of metal or stone which enter a hole in the upper side of one stone and also a hole in the lower side of the stone next above. They are used to prevent lateral motion between two stones. (Fig. 15.)

Cramps.—Metal bars having the ends turned at right angles to the body of the bar, which enter holes in the upper sides of adjacent stones. (Fig. 16.)

Quoin.—A corner-stone. A quoin is a header for one face, and a stretcher for the other. (Fig. 17.)

ART. 4.—CLASSIFICATION OF MASONRY.

11. As the term stone masonry includes properly all classes of construction in stone which require the employment of skilled mechanics or masons, any class of masonry

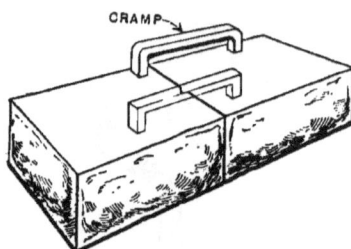

FIG. 16

may be laid dry, in lime mortar or in cement mortar, at will. On this point specifications should always be precise.

12. Rubble Masonry.—This is composed of unsquared stones. It is divided into: *Uncoursed Rubble* (Fig. 17), laid

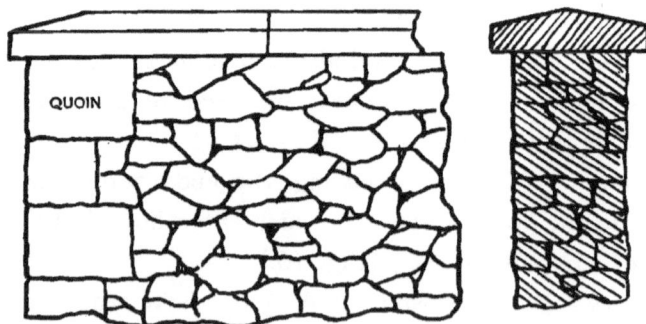

FIG. 17.

without any attempt at regular courses, and *Coursed Rubble* (Fig. 18), levelled off at specified heights to a horizontal surface. The stone may be required to be roughly shaped with the hammer, so as to fit approximately; it is then **called** squared rubble.

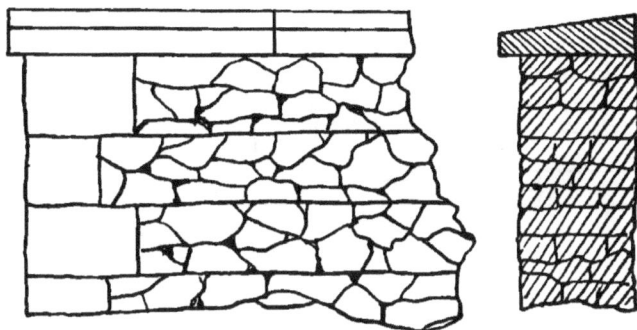

FIG. 18.

13. Squared-stone Masonry.—According to the character of the face, this is classified as *Quarry-faced* (**Fig. 19**) or as *Pitch-faced* (Fig. 20). If laid in regular courses of

FIGS. 19, 20.

FIG. 21.

about the same rise throughout, it is *Range*-work (**Fig. 21**). If laid in courses that are not continuous throughout the length of the wall, it is *Broken Range*-work (Fig. 22). **If not** laid in courses at all, it is *Random*-work (Fig. 23), and this is generally to be expected of this kind of masonry unless the specifications call for *Range*-work. This style of masonry is called *Block in Course* by the English. In quarry-faced and

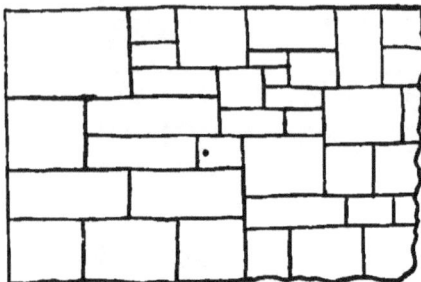 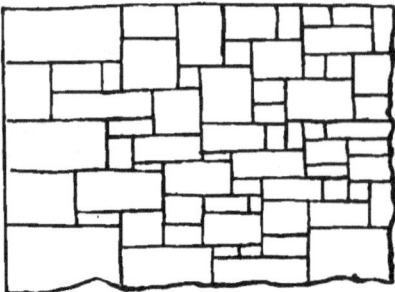

FIG. 22.   FIG. 23.

pitch-faced masonry quoins and the sides of openings are usually hammer-dressed. This consists in removing projections so as to secure a rough-smooth surface, and is done with the face-hammer, the plain axe, or the tooth-axe. This work is a necessity where door or window frames are inserted, and it greatly improves the general effect of the wall if used wherever a corner is turned.

14. Ashlar Masonry.—This is equivalent to "cut-stone masonry," or masonry composed of any of the various kinds

FIG. 24.

of cut stone mentioned above. As a rule the courses are continuous (Fig. 24), but sometimes they are broken by the introduction of smaller stones of the same kind, and then it is called *Broken Ashlar* (Fig. 25). If the stones are less than one foot in height, the term *Small Ashlar* is proper. The term *Rough Ashlar* is sometimes given to squared-stone masonry, either "quarry-faced" or "pitch-faced," when laid as *range*-work, but it is believed that it is more logical and more expressive to call such masonry "squared range-work." From its derivation, ashlar apparently means large square blocks, but practice seems to have made it synonymous with "cut stone," and this secondary meaning has been retained for convenience.

Dimension-stones are cut stones all of whose dimensions have been fixed in advance. If the specifications for ashlar

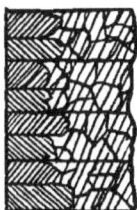

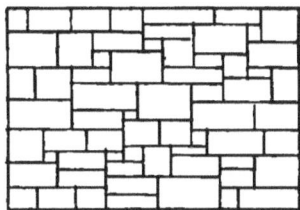

FIG. 25.

masonry are so written as to prescribe the dimensions to be used, it will not be necessary to make a new class of such stones.

Range-work, whether of squared stones or of ashlar, is usually backed up with rubble masonry, which in such cases is specified as Coursed Rubble.

Whatever terms are employed in common use to various classes of masonry, it is not safe to trust to them alone in preparing specifications for construction, but every specification should contain an accurate description of the character and quality of the work desired. Whenever practicable, samples of such kind of cutting and masonry should be prepared beforehand, and exhibited to the persons who propose to undertake the work.

Art. 5.—Strength of Masonry.

15. The strength of masonry will depend on the size of the blocks, the accuracy of the dressing, the bond, and the quality of the mortar.

16. Size of Stone.—The size of the blocks varies with the nature of the stone and quarry.

Some stones are strong enough to be used in any size, while others are limited by certain proportions of length, breadth, and thickness.

A common rule, for ordinary stone, is to make the breadth at least equal to the thickness, and seldom greater than twice this dimension, and to limit the length to within three times the thickness. When the breadth or the length is considerable in comparison with the thickness, there is danger that the block may break if any unequal settling or unequal pressure should take place. As to the absolute dimensions, the thickness is generally not less than one foot nor more than two; stones of this thickness with the relative dimensions just laid down will weigh from 1000 to 8000 lbs., allowing, on an average, 160 lbs. per cubic foot. Such a weight requires considerable effort of both men and machinery to set it in place.

17. Accuracy of Dressing.—Accurate dressing is essential for first-class work. If the individual blocks fit well on each other, the pressure is equalized, and cracking from this cause avoided. One of the objects of the mortar between the blocks is to equalize the pressure, and it will thus be seen that the less attention is paid to dressing the better the quality of the mortar should be. Many of the largest buildings of antiquity were erected without the use of mortar, but so well were the stones fitted together that the structures stood until overthrown by violence. Before finally placing a block of cut stone, a trial should be made to see that it fits properly.

18. Bond.—In stonework of any kind it is of the utmost importance to secure a good bond.

Headers should in all cases extend into the wall two thirds of its thickness, and preferably entirely through the wall from front to back. The number of headers in a given space varies much with the class of work. *Flemish* bond, consisting of alternate headers and stretchers, presents the best appearance and is probably as strong as any other plain bond. In foundation-work there should not be less than one header for every five square feet of surface of wall.

The vertical joints of the blocks of each course alternate with the vertical joints of the courses above and below it, or *break joints* with them. The vertical joints of one course are not to be less than four inches on one side of those in the next course, and the headers should rest as nearly as possible on the middle of the stretchers in the course below. In important work the proportion of headers, limiting sizes of stones, joints, etc., are always specified.

The mean thickness of a rubble wall should not be less than one sixth of the height; in the case of a dry stone wall the thickness should never be less than two feet.

The largest stones should be used for the foundation course.

Stratified stone should be laid on its *natural bed*, i.e., so that the direction of pressure comes at right angles to the direction of the laminæ.

CHAPTER II.

STONE-CUTTING AND MASONRY.

ART. 1.—DEFINITIONS.

19. STEREOTOMY, as applied to stone-cutting, is the art or science which teaches how to make drawings, patterns, bevels, etc., by which the mason may shape blocks of stone which when fitted together form a predetermined whole.

20. The stone structures treated of in these limits may be divided into two general classes:

1st, those having plane surfaces only; and, 2d, those having plane and curved surfaces. The second class may be variously subdivided as structures with developable surfaces, with warped surfaces, and with surfaces of double curvature.

In engineering work structures of the first class predominate, and in architectural work those of the second. Examples of the first class may be found in retaining-walls, wing-walls, culverts, piers, canal-locks, buttresses, and fortifications; and of the second class in piers, arches, vaults, domes, towers, aqueducts, etc. Numerous engineering structures and buildings show examples of both classes. A sharp division is therefore neither necessary nor desirable, and no effort has been made in these pages to keep them separate, except in the theoretic examples.

21. The first operation in all stone-cutting is to form the joints of one surface, and from these all the other bounding

18

planes of the finished stone are derived (Fig. 26). A line is first drawn around the stone and the joints either pitched off to this line with the pitching-chisel, as in Fig. 20, or a chisel-draft is sunk all around the face until the line is reached (Fig. 27). The rough projections of the central portion are then

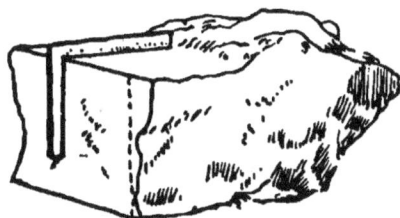 

FIG. 26.   FIG. 27.

brought to the plane of the margin by means of chisel and crandall, and finished according to specifications.

22. Directing-instruments.—These are of three kinds.—*patterns, templets,* and *bevels,*—and must of course all be full size.

Patterns show the forms of plane or of developable surfaces, and in the latter case are made of flexible material.

Templets give the forms of required edges or other distinguishing lines of a surface. A pattern is used to lay out the *relative* position of the edges, whereas a templet defines any particular edge or edges.

Bevels show the diedral angles between surfaces. The common carpenter's square is a special form of bevel, and the straight-edge a special form of templet.

Thus in Fig. 28 patterns would give the outline of the faces marked *A, B,* and *C,* and a bevel or templet, or both, might be used for *D.*

23. It is the place of the engineer or architect to supply the stone-cutters with such drawings as will enable them to make all the necessary directing-instruments. In practical

work it is therefore not customary to make such detailed
drawings as shown here for some of the simpler examples;
but in order that the student may know to what extent to go

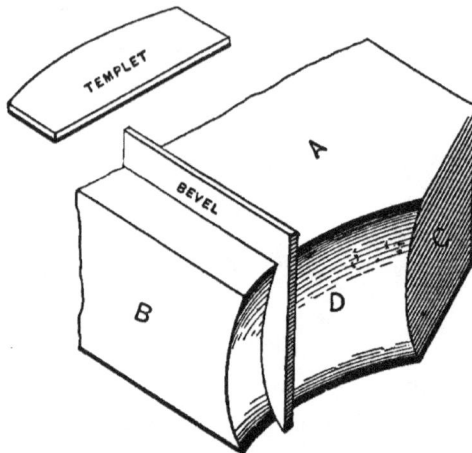

FIG. 28.

into detail he must himself know how to make these details
from the drawings.

The plates, which we will now discuss, fully illustrate
this point.

PLATE 1

GOTHIC BUTTRESS

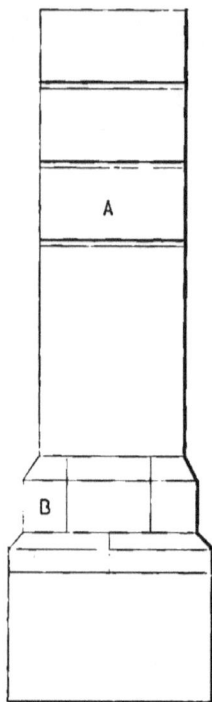

FRONT ELEVATION

SIDE ELEVATION

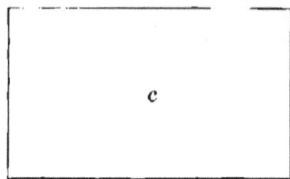

c

PATTERNS OF
STONE AT A

b

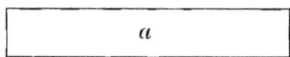

a

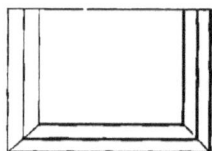

PLAN AT BASE

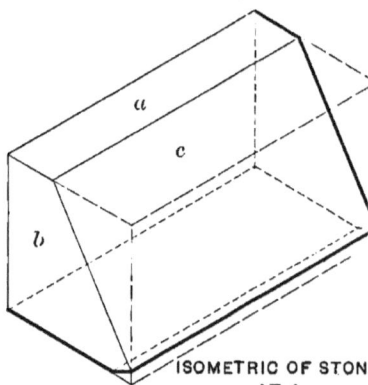

ISOMETRIC OF STON
AT A

PLAN AT CAP

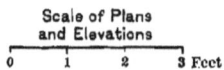

Scale of Plans
and Elevations

0   1   2   3 Feet

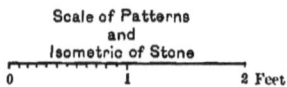

Scale of Patterns
and
Isometric of Stone

0   1   2 Feet

ART. 2.—GOTHIC BUTTRESS.

PLATE I.

24. Plate I illustrates a simple Gothic buttress, examples of which may be seen on the walls of many churches.

Since the corresponding points on the various figures bear the same letters, the student should have no difficulty in reading the drawing. For the sake of clearness a horizontal section at the base is shown, and another just below the cap. The stone A, of which the details are drawn, is termed a *kneeler*, and the manner in which it serves to support the coping, or cap, is evident. The details show an isometric view of the stone A, and patterns of the different faces. The size of the stone in the rough is indicated by the broken line on the isometric. In practice these patterns are made full size and cut out of paper, cardboard, or zinc.

25. The method of working in this case would be about as follows: Having cut the rough block of stone to a prism of nearly the desired dimensions, as explained above (§ 21), the face a is marked out from its pattern and brought to a plane; and the back of the stone and the two faces marked b worked at right angles to a.

The surfaces b are next marked, and c roughly cut into shape; b is dressed, drafts run around the edges of c, and its surfaces finally worked. The bottom is now easily dressed to the proper plane, and the stone is ready to hoist into place.

26. For a working drawing the details shown on this plate would be omitted, because the foreman of the masons could readily make the patterns himself from the general drawings for such a simple case.

27. Exercises.—1. Draw an isometric view and the necessary patterns of the stone *B*.

2. Draw a buttress similar to the one shown on Plate I against a wall 3 ft. thick at the base, 13 ft. high, batter of face of wall 1 inch per foot, height and dimensions of base of buttress as in illustration, batter of its front 1 inch per foot and its sides ¾ inch per foot. Slope of top and water-tables as shown. Make the necessary details for the stones similar to *A* and *B*.

3. Take the measurements and make drawings of a buttress on some building.

PLATE 2
PERSPECTIVE OF ARCH

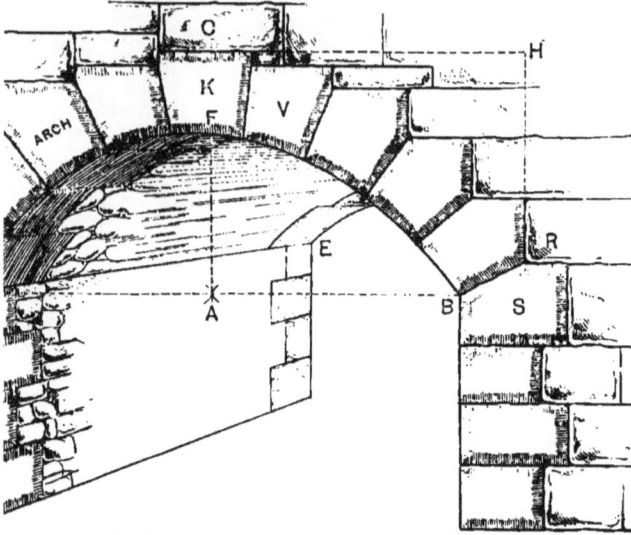

NAMES OF PARTS.

D A B = Span
D E = Springing Line
V - Voussoirs
K = Central Voussoir or
    Keystone
M C R - Extrados
C = Crown

A F = Rise
S = Springer
D F B = Intrados
R H C = Spandrel
C R = Haunch
B R = Skewback

PLATE II.

28. This plate clearly illustrates and defines the various parts of an arch.   Additional definitions are:

Soffit : the concave surface of the arch.

Back : the convex surface of the arch.

Spandrel-filling : the filling in the space *CHR*.

All the definitions given for one half the arch naturally apply to the other half; thus there are two haunches, two skew-backs, etc.

String-course : a course running the length of the arch, parallel to *DE* in a right arch.

Ring-course : a course parallel to the face of the arch.

Coursing-joint : a joint between adjoining string-courses. Also called *Bed-joint*.

Heading-joint : a joint between adjoining ring-courses.

Arch-sheeting : all of the arch masonry of the arch proper, except the face-voussoirs.

The face of the arch is also called *The Head*, and the face voussoirs *Ring-stones*.

PLATE III.

29. On Plate III are shown methods of constructing the principal styles of arches used in architecture and engineering.

The style of the arch to be employed is governed by the style of architecture employed, the purpose it is to serve, and the local conditions. The full-centred or semicircular arch (Fig. 1) is the strongest, but is often replaced by the segmental (Fig. 6), owing to limited space for the rise. The elliptic arch (Fig. 4) is generally considered the most graceful; and the Gothic and Tudor arches (Figs. 3 and 5) are employed chiefly in certain styles of architecture.

The Gothic or pointed arch, shown in Fig. 3, is drawn by striking intersecting arcs from a, a as centres, with the span as radius.

30. If the intrados is to be an oval and its rise is to be not less than one third the span, a three-centred oval will generally give a curve of form more pleasing to the eye than one of a greater number of centres.

If we assume the radius of the curve at the springing-lines, the general construction of a three-centred oval is as follows:

On the span and rise set off at AD and EE a distance less than the rise HF. Draw DE, bisect it by a perpendicular, and extend this perpendicular to meet FH produced in C. Then will C and D be two of the required centres.

An infinite number of ovals may be thus constructed for the same span and rise, and a third condition may be imposed

PLATE 3
MASONRY ARCHES

SCALE OF FEET
0 1 2 3 4 5 6

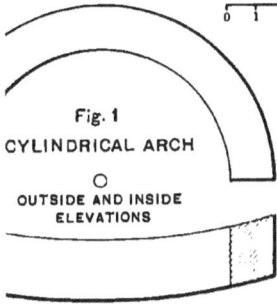

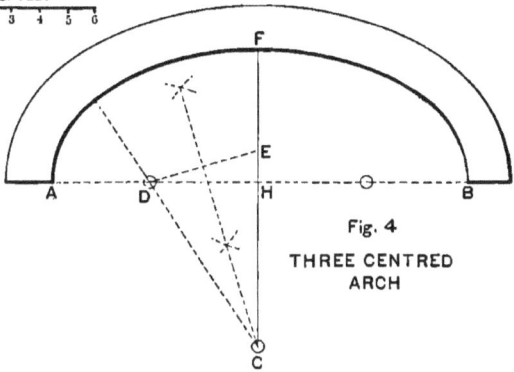

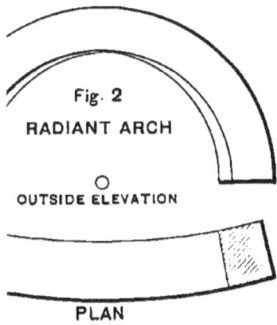

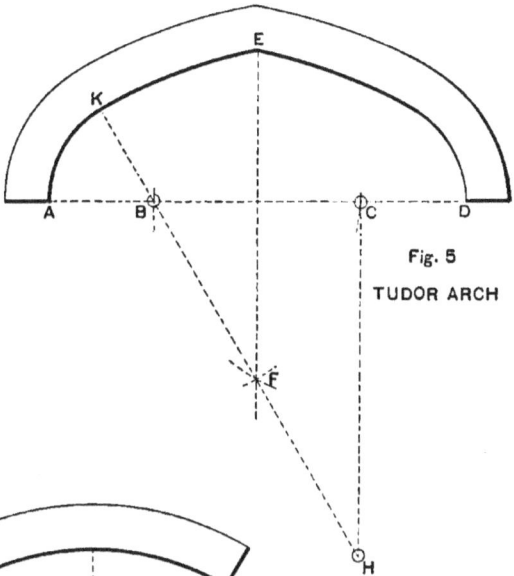

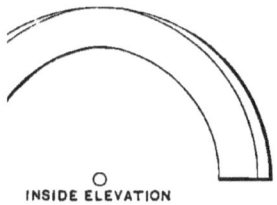

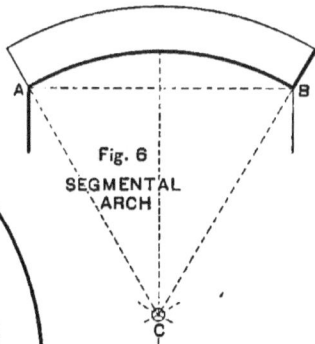

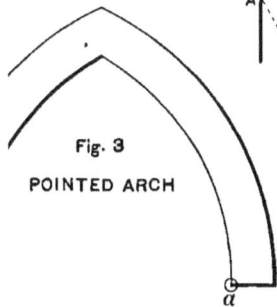

Fig. 1
CYLINDRICAL ARCH

OUTSIDE AND INSIDE
ELEVATIONS

Fig. 4
THREE CENTRED
ARCH

Fig. 2
RADIANT ARCH

OUTSIDE ELEVATION

PLAN

Fig. 5
TUDOR ARCH

INSIDE ELEVATION

Fig. 6
SEGMENTAL
ARCH

Fig. 3
POINTED ARCH

if it be desired to make a determinate solution. (See Wheeler's "Civil Engineering," p. 258.)

31. The four-centred Tudor arch (Fig. 5) may be constructed as follows: Divide the span *AD* into four equal parts, *AB*, *BC*, etc. From *B* and *C* as centres, with radius equal to *BC*, describe arcs intersecting in *F*. Draw *BF* and extend it to meet a perpendicular to *AD* through *C*. With *B* as centre and radius *AB* describe *AK*, and with *H* as centre and radius *HK* describe *KE*. Similarly for the other half of the arch.

The segmental arch (Fig. 6), such as is used to span openings over doors and windows, is constructed on the equilateral triangle *ABC*, having a side equal to the span.

PLATE IV.

32. The Recessed Flat Arch, or Plate-band.—This form of arch is used where it is desirable to have the top of the opening present a flat surface. The bearing power of this construction is small, and any considerable weight above it is borne by a *relieving-arch* or a beam. The lintels over the doors and windows of a brick building are simple examples. This arch should not be used for spans exceeding ten feet.

Having given the necessary dimensions, the front elevation and top view are drawn. A top view is shown in place of the usual plan in order to bring out the joints of the arch. Thus the line *f–g* shows a joint on the back of the arch; *h–i* on the elevation corresponds to *h–i* on the top view. There being a shoulder at *i* which does not appear on the elevation, and the line *i–l* being in a vertical plane, the points cannot be lettered alike without tending to confuse. The student should draw a vertical section through the keystone, which will materially aid him to a better understanding of the structure.

The isometric of the stone *A* may be drawn as follows: From any convenient point, as *O*, draw the usual co-ordinate axes, *O–a*, *O–2*, and *O–13*. Lay off *O–2* equal to the width of the stone, equal to 12–2 on the elevation. Draw the front of the stone, all of whose lines except 4–5 show in their true size, and may therefore be taken from the elevation. Lay off 1–14 and 14–15, equal to 1–14 and 14–8 respectively on the top view, and draw 1–15. 15–16 = 8–9, 16–12 = 15–12, and 12–13 = 10–13. As a check *O*–13 should equal *n–O*.

PLATE 4
RECESSED FLAT ARCH

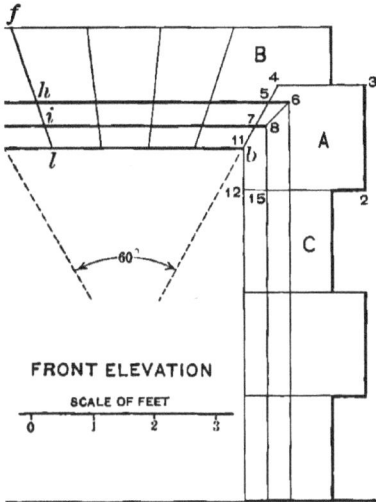

FRONT ELEVATION

SCALE OF FEET

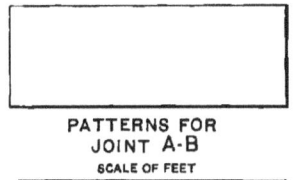

PATTERNS FOR
JOINT A-B

SCALE OF FEET

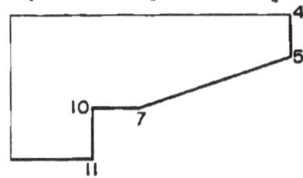

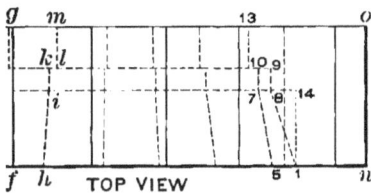

TOP VIEW

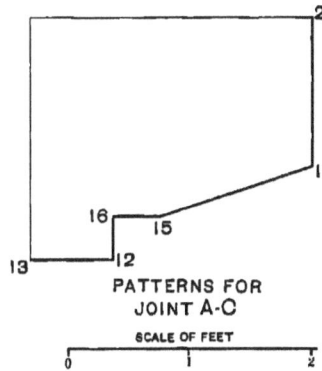

PATTERNS FOR
JOINT A-C

SCALE OF FEET

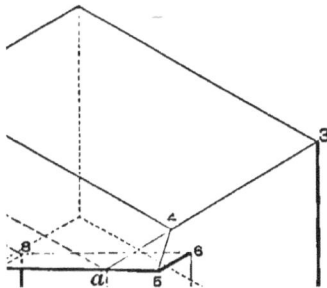

ISOMETRIC OF
STONE AT A

SCALE OF FEET

From the points 13, 12, 16, and 15 draw the verticals as taken from the elevation, and the drawing is readily completed. The patterns, being numbered to correspond to the numbering on the other figures, need no further explanation.

33. Exercises on Plate IV.—1. Draw a vertical section through the keystone.

2. Draw an isometric of, and all necessary patterns for, the keystone and the stone *B*.

3. Draw a recessed circular or pointed arched gateway of dimensions similar to the flat one shown; adding three or four stone steps to the front.

PLATE V.

34. Arches in Circular Walls.—An arch in a circular wall is a construction sometimes met in architectural and in engineering work.

It is generally avoided, owing to the difficulty of constructing it, and the not pleasing effect of the finished structure. There are instances, however, when it is necessary to use it, and an explanation of a simple case is inserted here. For the more complex cases the student is referred to Émile Lejeune's " Traité Pratique de la Coupe des Pierres."

35. Almost any form of arch may be constructed in a circular wall. Two general divisions may be noted—the cylindrical arch, and the radiant arch. The first is the ordinary arch with cylindrical intrados and extrados, whereas in the second the elements of intrados and extrados radiate from a common vertical axis. The elements are, however, in every case parallel to the springing-plane—a point which should be carefully noted, and which will readily appear from an inspection of Plate V. The outer and inner openings of the cylindrical arch are the same in size, but in the radiant arch the inner opening is smaller.

36. A cylindrical arch in a circular wall may be considered as a case of two cylinders intersecting at right angles, the axis of the one being vertical and that of the other horizontal. We will explain how to obtain all the necessary patterns, bevels, etc., have given plan, elevation, etc., of the arch.

37. Fig. 1 shows the elevation of a cylindrical circular arch and a portion of the circular wall containing it. The plan

PLATE 5

CYLINDRICAL ARCH
IN A
CIRCULAR WALL

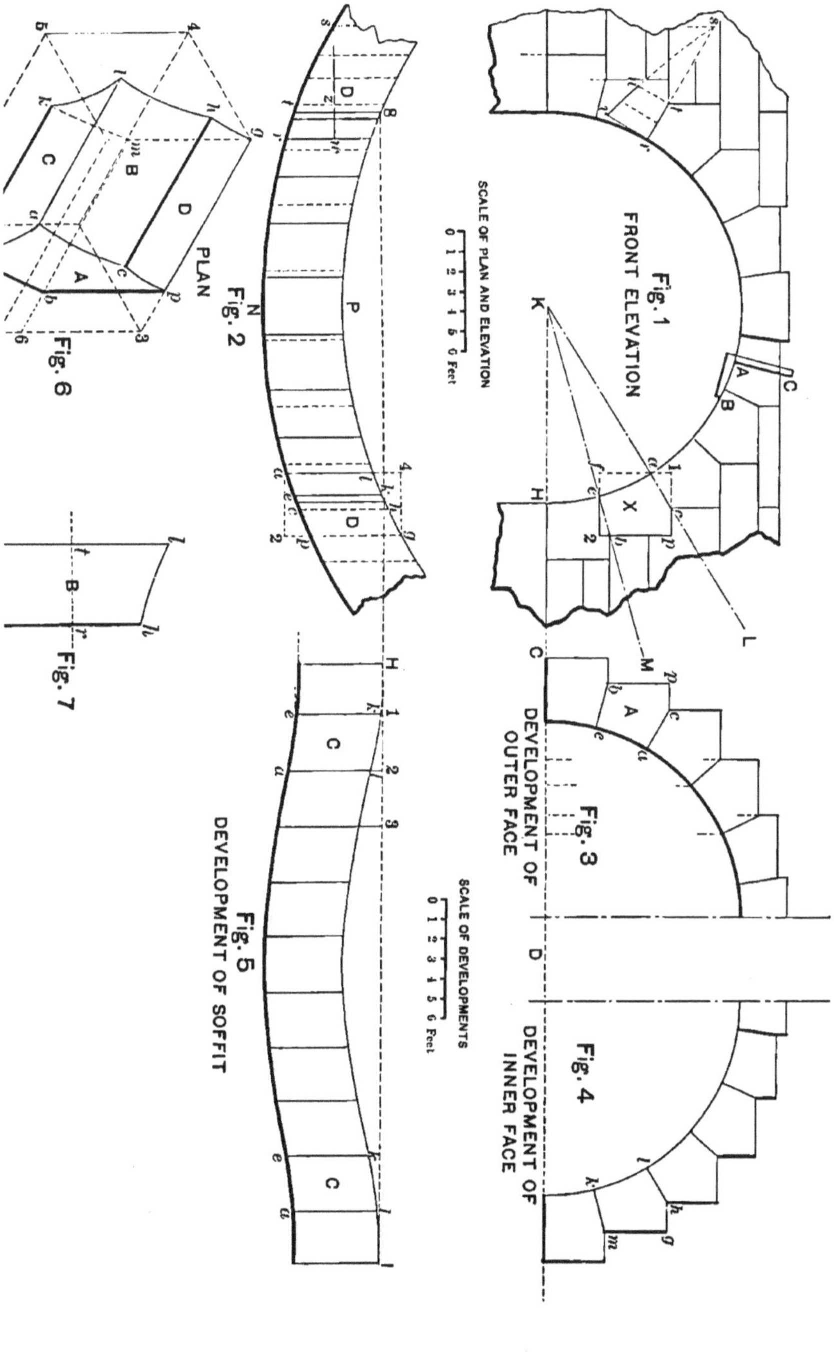

Fig. 1
FRONT ELEVATION

Fig. 2

PLAN
Fig. 6

Fig. 7

SCALE OF PLAN AND ELEVATION
0 1 2 3 4 5 6 Feet

Fig. 3
DEVELOPMENT OF
OUTER FACE

Fig. 4
DEVELOPMENT OF
INNER FACE

Fig. 5
DEVELOPMENT OF SOFFIT

SCALE OF DEVELOPMENTS
0 1 2 3 4 5 6 Feet

shows the joints as though the soffit were viewed from below, which is done to show its joints more plainly. Let us take some one stone, as *X*, and show how to derive all the patterns necessary to work its faces.

Draw first an isometric view, as Fig. 6, in order to obtain a better conception of its appearance. Briefly, this may be constructed thus: Around the elevation and plan of *X* draw rectangles showing the limits of the rough stone from which to work the finished block. Draw an isometric block with these dimensions, as *f*–1–3–2, etc., and on this plot the corners of the stone—remembering that three such co-ordinates are necessary to locate a point not in an isometric plane. Thus the point *b* requires the co-ordinates *f*–2 and 2–*b* from Fig. 1 and 2–*p* from Fig. 2; which are laid off as *f*–2, 2–6, and 6–*b* respectively in Fig. 6.

The face *A*, having the curve of the wall as shown by the lines *c*–*p* and *e*–*p*, Fig. 2, must be developed to show its true size. The developments of one half the outer and one half the inner faces is shown in Figs. 3 and 4. Set off along the line *C*–*D*, Fig. 3, the outer curve of the wall, by points from Fig. 2; erect ordinates at these points and project the proper heights over from Fig. 1. (See Descriptive Geometry for development of cylinder.) The pattern is now shown in its true size at *b*–*c*–*a*–*c*–*p*, Fig. 3. The corresponding one for the inner face is found on Fig. 4.

The pattern for *B* requires us to find the true size and appearance of the radial joints *a*–*c* and *e*–*b*. These joints are formed by the intersection of the planes *K*–*L*, *K*–*M*, etc., perpendicular to the vertical plane of projection, with the vertical cylinder or portion of cylinder forming the wall. They are therefore arcs of ellipses, and may be shown in their true size by revolving parallel to one of the planes of projection or constructing the ellipse, having given its axes. The former method is shown here, and for clearness the corresponding stone on the left of the arch is taken. The chord *s–r*, Fig. 1, is used as an axis. When the ellipse is revolved into *V*, the point *s* remains stationary, being where the axis intersects the arc; *r* falls at *v*, a distance equal to *wr*, *t* at *u*, and other points similarly. *u–v* shows the arc we want, and Fig. 7 gives the construction of the pattern—the lines *c–h* and *a–l* being obtained from Fig. 2.

To construct the pattern for *C* we must develop the under surface or soffit of the arch. If the arch were in a straight wall, its development would of course be simply a rectangular band equal in length to the semicircumference of the circle of radius *K–a* and in width to the thickness *N–P*. But owing to the circular shape of the wall the development departs from this rectangular form, as will be shown if a chord be drawn from *h* to 8, Fig. 2. To develop, then, lay off, by points, on the indefinite line *H–I*, Fig. 5, the length of the arch from Fig. 1, as *H–*1, 1–2, etc., equal to *H–c*, *e–a*, etc., respectively, in Fig. 1; draw perpendiculars at these points, and project the points *k*, *l*, etc., over from Fig. 2. Through the points thus found draw the curve of the developed intrados. *C* shows the desired pattern. *D* is shown in its true size in Fig. 2.

38. At *CAB*, Fig. 1, is shown an arch square. It is a bevel, the arm *AC* of which radiates from the centre of the arch, and the arm *AB* is curved to the curve of the intrados, as shown on the elevation.

39. The stone *X* may now be worked as follows: The face *bpgm*, Fig. 6, is rectangular, and needs no further explanation. The face *D* at right angles to this one is also readily worked, its pattern being shown on Fig. 2. A bevel, having the angle *acp*, Fig. 1, is made, and the face *B* worked by means of it and the pattern, Fig. 7. The soffit *C* is now worked by means of the arch square and the pattern *calk*, Fig. 5, this pattern being made of zinc, and bent to the proper curve on the arch square. The development *A*, Fig. 3, bent to the curve *ap*, Fig. 2, will give the pattern of *A*, Fig. 6, and therefore a templet for *eb*. A draft *eb* having been sunk, the face *A* is readily worked. The other faces are obtained similarly.

40. Exercises.—1. Draw an isometric and construct all necessary patterns and bevels for the keystone.

2. Replace the arch in the illustration by a three-centred one and make the drawings.

3. Make the drawings when the face of the arch is moulded.

ART. 4.—INTERSECTING ARCHES.

PLATE VI.

41. We will now deal briefly with the surfaces formed by the intersection of two arches.

Whenever two intersecting surfaces of the second degree (ellipse, hyperbola, parabola) have one diameter in common, their intersections lie in planes, and are consequently projected on the plane of the axes as straight lines. Two cylindrical arches having each the same rise (the spans may or may not be equal) will intersect in ellipses, and this is true whether the intersection be right or oblique. The curves thus formed by the intersection of two vaults are called groins, or groin curves.

Two general divisions of intersecting arches may be noted, forming, 1st, *The Cloistered Vault*, and, 2d, *The Groined Vault*.

The cloistered vault (Figs. 4 and 5) is formed when the arches *meet* to form a covering for a closed area, making a vaulted ceiling for a room. The groins in this case form valleys between the vaulting surfaces. The effect may be seen in any room which has a coved ceiling, i.e., one in which the walls and ceiling are connected by a curve instead of a right angle.

The groined vault (Figs. 1, 2, and 3) is formed when the arches *cross* each other, forming the vaulted passage called cross-vault (German *Kreuzgang*). Fig. 1 shows plainly how the groin curves, as *fbg*, are formed by the intersection of the

PLATE 6
GROINED
AND
CLOISTERED VAULTS

SCALE OF FEET

0 1 2 3 4 5 6 7 8 9 10

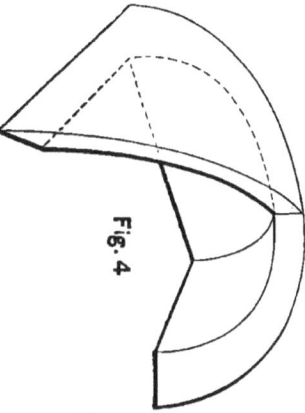

Fig. 4

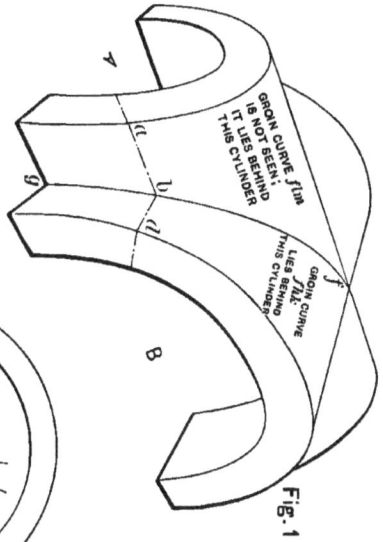

Fig. 2

GROIN CURVE *j'lm*
IS NOT SEEN:
IT LIES BEHIND
THIS CYLINDER

GROIN CURVE *j'kz*
LIES BEHIND
THIS CYLINDER

B

Fig. 1

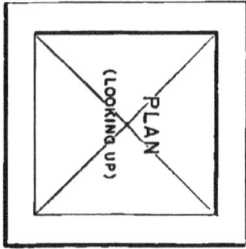

PLAN
(LOOKING UP)

SECTION

SECTION

Fig. 5

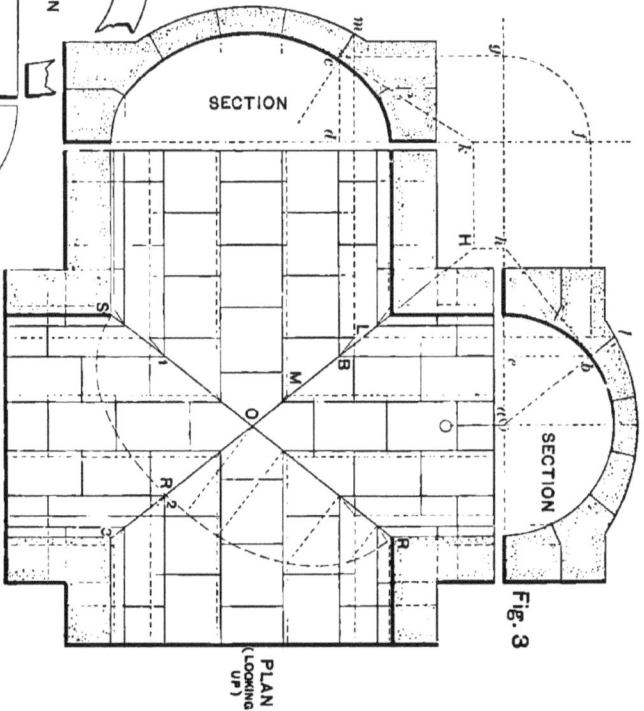

SECTION

SECTION

PLAN
(LOOKING UP)

Fig. 3

different elements, *ab*, *db*, etc. It will be seen that the groins
in this case project into the passage—just the opposite from
the cloistered groin. If in Fig. 1 the elements of the cylinder
A lay between the groins *fbg* and *fhk* and the corresponding
ones to the left, and the elements of the cylinder *B* lay between the groins *fbg* and *flm* and the corresponding ones to
the rear, the passageway would be closed and the groined
vault would become a cloistered vault.

Fig. 2 shows a groined vault from below.

42. The Right Groined Vault.—Referring now to Fig.
3, let the two vaults be given as in the plan, intersecting at
right angles. The joint lines of the intrados are, as in Plate
V, shown as full lines, because they are of more importance
than those of the extrados, which latter is frequently left
unfinished. Let the right section of the one vault be a circle
of radius *a–b*, radius of extrados being equal to depth of
keystone plus ¾ diameter of arch. This increase of radius
gives a proper thickness towards the springing of the arches.
Now let us impose the conditions that corresponding points
on the two arches be the same height above the springing-
plane, and that all bed-joints be normal to the soffits.
The joints of the circular arch—call it No. 1—will of course
radiate towards the centre *a* of the intrados, while those of
the other arch, No. 2, will be determined as follows:

First, to find the curve of the intrados of No. 2. Project
the point *b* of No. 1 on the groin, in *B*; from *B* draw *Bc*
parallel to the axis of No. 2, and lay off *dc* equal to *eb*; or
project the point as shown by the dotted construction *bfgc*.
Sufficient points having been thus found, the ellipse is readily
drawn.

To draw the joints normal to the intrados we must first
draw tangents at the points just determined. Draw a tangent

bh in No. 1 and project it on the groin as *BH*; draw *HK* parallel to the axis of No. 2, and join *k* and *c* for the new tangent. The normal is, of course, drawn perpendicular to this tangent, and the point *m* fixed by laying off its height above the springing-plane, as shown by *l* in No. 1. The reason for this construction will be evident if the student remember that the tangent to the intersection of two surfaces lies in the intersection of the tangent planes.

PRS shows the groin revolved into *H*, about *POS* as an axis.

Joining the points *LMO*, etc., gives the groin of the extrados.

43. Exercises.—1. Isometric and patterns of keystone and a springing-stone.

2. Two vaults of unequal spans uniting at an angle of 60°.

3. Details for 2.

4. Let vault No. 1 stop on the line *S*–3 and No. 2 on the line *P*–3.

PLATE 7
6 FOOT ARCH CULVERT
AUBURN & ITHACA R. R.
3RD CLASS MASONRY

LONGITUDINAL SECTION ON AXIS OF ARCH

½ PLAN OF ENDS AND EXCAVATION

SCALE OF FEET
0 1 2 3 4 5

ARCH HAS ⅜ JOINTS

NATURAL SURFACE

½ END VIEW AND SECTION A-B

LOCATION PLAN SHOWING ELEVATIONS

SCALE OF FEET

ART. 5.—ARCH-CULVERTS.

PLATE VII.—6-FT. ARCH-CULVERT, AUBURN & ITHACA
R. R.

44. Complete working drawings are shown on this plate of
a small culvert built of rough masonry.

45. Third-class masonry, as used on this road, is specified
as being rubble-work (§ 12), laid with mortar in irregular
courses, and will consist of stone containing generally six
cubic feet each, so disposed as to make a firm and compact
work; and no stone in the work shall contain less than three
cubic feet, except for filling up the interstices between the
large blocks in the heart of the wall. At least one fifth of the
face shall be composed of headers extending full size four feet
into the wall, and from the back the same proportion and of
the same dimensions, so arranged that a header in the back
shall be between two headers in the face. The corner-stones
shall be neatly hammer-dressed, so as to have horizontal beds
and vertical joints. Also, drafts to be cut on all covers of
masonry.

46. The stones for the faces of the arch, coping, etc., are
fully figured, and the mason is therefore able to cut them
without further detailed drawings.

The method adopted on this plate of showing one half
views is done only to economize space. It is not recom-
mended for working drawings of this class. The little extra
time required to make the complete drawing is more than

balanced by the time saved during construction in avoiding numberless questions and the probability of making corrections on work done.

This drawing is so well figured and lettered throughout that the student needs only to study it carefully in order to fully understand it.

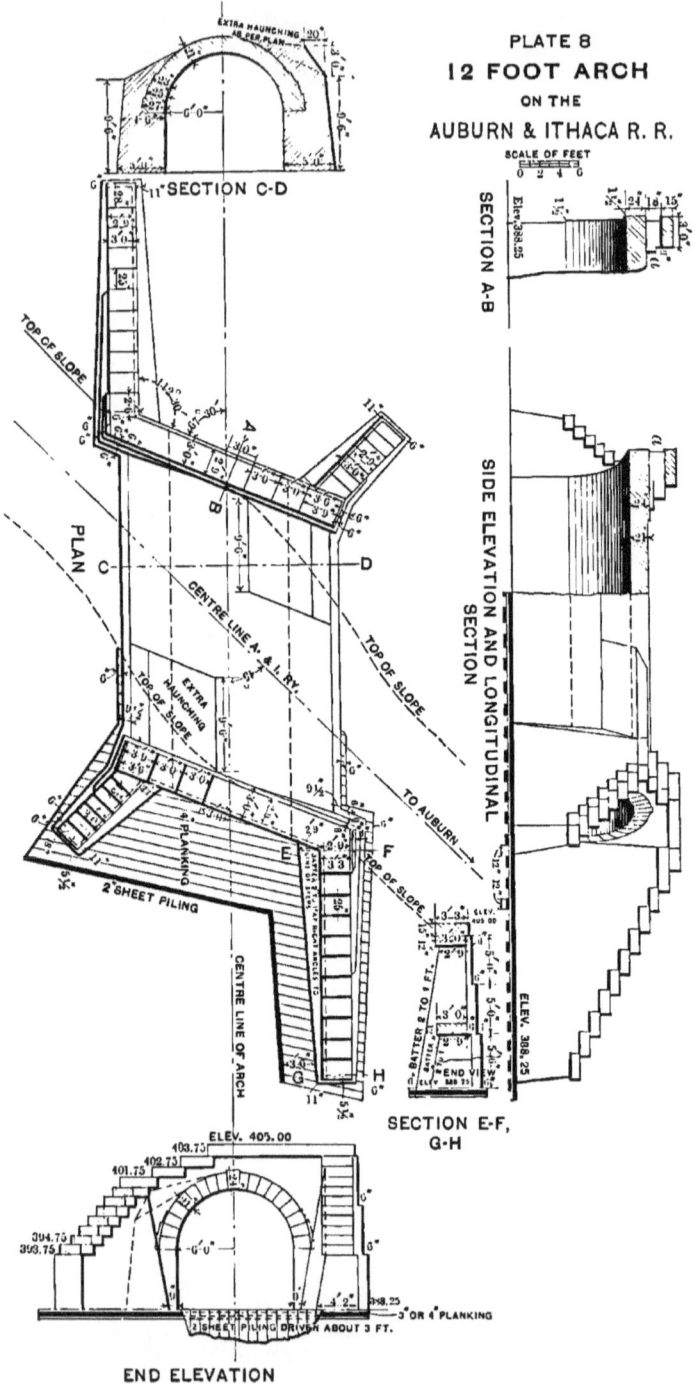

PLATE 8

12 FOOT ARCH

ON THE

AUBURN & ITHACA R. R.

SCALE OF FEET

SECTION A-B

SECTION C-D

EXTRA HAUNCHING AS PER PLAN

TOP OF SLOPE

PLAN

CENTRE LINE A. & I. RY.

EXTRA HAUNCHING

TOP OF SLOPE

TOP OF SLOPE

TO AUBURN

SIDE ELEVATION AND LONGITUDINAL SECTION

2" SHEET PILING

PLANKING

CENTRE LINE OF ARCH

SECTION E-F, G-H

BATTER 8 TO 1 FT.

END VIEW

ELEV. 398.25

ELEV. 405.00

403.75
402.75
401.75

394.75
393.75

388.25

3" OR 4" PLANKING
2" SHEET PILING DRIVEN ABOUT 3 FT.

END ELEVATION

PLATE VIII.—12-FT. ARCH ON THE AUBURN & ITHACA R. R.

47. This is a good illustration of a case where the railroad does not cross the stream at right angles. To save
building a larger structure than necessary recourse is had to
a skew-face arch, i.e., one in
which the axis is oblique to
the plane of the faces. Fig.
28 illustrates this, the dotted
lines showing where the faces
of a plain right arch giving
the same width of roadway
as the skew-face arch would
come.

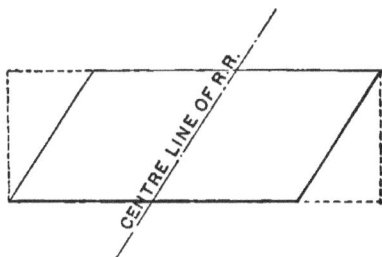

Fig. 28.

The extra haunching shown on the various figures is to
strengthen the arch, and the planking, held by heavy sheet
piling at the ends, serves to prevent the stream from scouring
under the arch. The wing-walls are built with an even batter on the inside faces, while for the outside faces and back of
the parapet-walls the same effect is obtained by building
steps. This is plainly shown on the right wing-wall of the
end elevation.

The right section *CD* shows the dimensions of the abut-
ment-walls and the increased depth of the sheeting towards
the springing.

The student should note that, although the end elevation
shows the arch as a semicircle, it is really a semiellipse with
diameters equal to 13 feet and 12 feet respectively.

Stones such as *a* on section *AB* and on longitudinal

section and in other places are not filled in with cross-section
lines, because it is desired to bring the cut-stone work into
prominence.   The general arrangement of the drawings on
this plate is very good, the draughtsman being enabled to
readily project from plan to elevation, sections, etc.

48. In passing it may be well to note the difference existing between the term plan as used by the engineer and by
the architect.   The first uses it in the strictly correct sense,
i.e., horizontal projection, while with the second it generally
means a horizontal section at some point; thus the plan on
this plate would be an architect's top view.

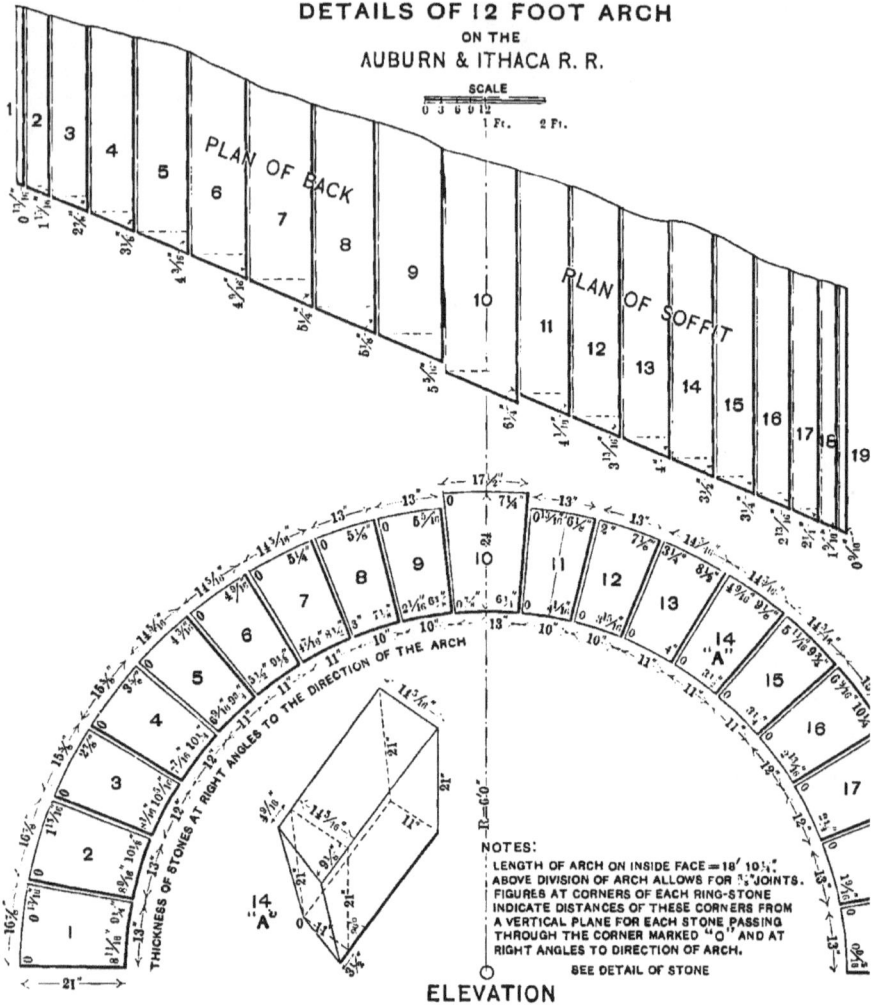

PLATE 9

DETAILS OF 12 FOOT ARCH
ON THE
AUBURN & ITHACA R. R.

SCALE
0 3 6 12
1 Ft.   2 Ft.

PLAN OF BACK

PLAN OF SOFFIT

THICKNESS OF STONES AT RIGHT ANGLES TO THE DIRECTION OF THE ARCH

R = 6'0"

14 "A"

NOTES:
LENGTH OF ARCH ON INSIDE FACE = 18' 10¼".
ABOVE DIVISION OF ARCH ALLOWS FOR ⅜ JOINTS.
FIGURES AT CORNERS OF EACH RING-STONE
INDICATE DISTANCES OF THESE CORNERS FROM
A VERTICAL PLANE FOR EACH STONE PASSING
THROUGH THE CORNER MARKED "O" AND AT
RIGHT ANGLES TO DIRECTION OF ARCH.

SEE DETAIL OF STONE

ELEVATION

PLATE IX.—DETAILS OF ARCH ON PLATE VIII.

49. Here are shown in a complete and simple manner the necessary details for the face-voussoirs of the skew-face arch on Plate VIII. The face is projected on a vertical plane at right angles to the axis of the arch, thus showing a semicircle instead of a semiellipse. The 18′ 10¼″ given in the notes as the length of arch on inside face is the length of the intrados of a right section. As the thicknesses of the archstones are measured in this plane, it is evident that if proper measurements are given from such a plane to the outside corners of each face-stone the face of the arch will have the desired skew. These distances may be readily obtained from the complete plan of soffit and back. The plan as given shows the method, and for the sake of clearness the soffit and back are shown separate. Fig. 29 shows both projected on the same figure.

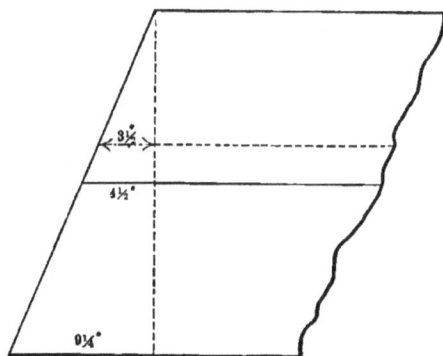

FIG. 29.

The dimensions given on the drawings, together with the specifications for the kinds of masonry of which the structure is to be built, will enable the mason to shape the stones properly.

Masonry for arch-culverts is sometimes divided into first and second class, but a general specification is as follows:

50. Arch-culvert Masonry.—The abutments are to be built of second-class masonry (§ 54).

The beds and joints of the arch-stone are to be accurately cut, and to be laid in courses throughout. The ring-stone will be neatly cut, and composed of alternate long and short bond-stones of not less than three feet and eighteen inches respectively. The parapet and wing walls will be built similarly to the abutments, and surmounted with a well-dressed coping, not less than fourteen inches thick and three feet wide, to be paid for at the same rate as for arch-culvert ma-sonry. The outside stones to be laid in cement mortar, and the whole wall to be thoroughly grouted, each course separately. The spandrel-backing (haunching) to be good rubble-work, laid in mortar, built as directed by the engineer, and to be paid for at the same rate as third-class bridge masonry.

In any arch-culvert or bridge the abutments, wing and spandrel walls, and backing, or any of these, may, at the discretion of the engineer, be built of rubble-work, similar to that of third-class bridge masonry (§ 42), and when so built shall be paid for at the same price as that class of masonry.

51. Exercises on Plates VIII and IX.—1. Make an end elevation taken at right angles to the plane of the face of the arch.

2. Draw isometrics of, and patterns for, some of the coping-stones of the wing and parapet walls.

3. Check the corner dimensions given for the keystone on Plate IX, and make an isometric drawing of the stone.

PLATE 10

SOUTH ABUTMENT & PIER

JORDAN CREEK BRIDGE

LEHIGH VALLEY R. R.

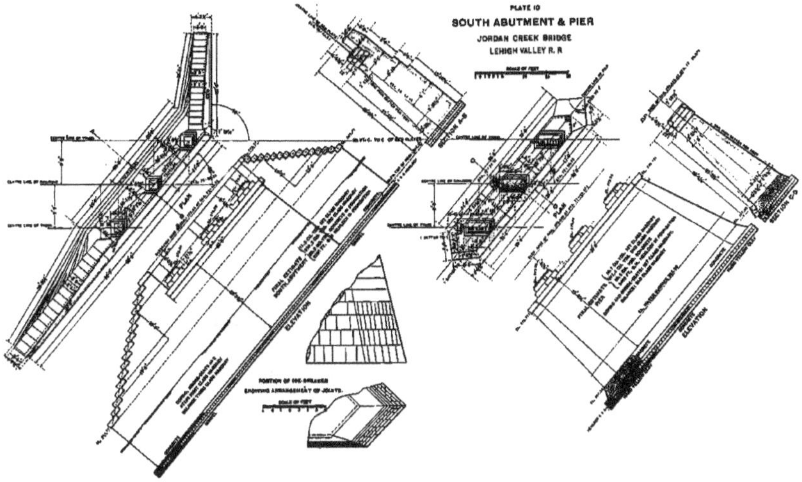

Back of
Foldout
Not Imaged

ART. 6.—RAILROAD-BRIDGE MASONRY.

PLATE X.—ABUTMENT AND PIER ON LEHIGH VALLEY RAILROAD.

52. The drawings for this piece of work are much more nished in appearance than the actual structures. By comarison of this plate with No. 11 it will at once be seen that lthough a great many dimensions are given for the abutment nd pier there are but very few dimension-stones and no deails for any cut stone. A reference to the specifications for ne kinds of masonry used (§§ 50, 53, and 54) shows plainly nat no special details are needed. The original drawings re unusually well figured, but for the purpose of reproducion it was found necessary to omit some. The student's ttention is, however, called to the fact that it is always well o give full dimensions for a working drawing if it is exected to carry out the work in exact accordance with them. icaled dimensions, especially from a blue print, are apt to be ery misleading. A close study of working drawings and oting carefully the work being done on similar structures in he course of erection will now teach the draughtsman where imensions are most needed and where he can save himself nnecessary work.

Here, as on Plate VIII, the faces of the different walls ave an even batter, while the backs are stepped.

Notice how the steps on the wing-walls radiate from the teps on the back of the abutment. The sections *AB* and *D* are taken on broken lines in order that the thickness **of** all of the stones supporting the bed-plates may be shown on one section.

The ends of the pier are shaped as shown to turn aside driftwood, ice, etc., and to prevent the forming of eddies around the pier, which might scour at the foundation.

53. First-class Masonry will be first-class rock range-work (§ 14). The stone to be accurately squared, jointed, and bedded, and laid in courses not less than ten inches thick, nor exceeding twenty-four inches in thickness, regularly decreasing from bottom to top of pier or abutment. The stretchers shall in no case have less than sixteen inches bed for a ten-inch course, and for all courses above sixteen inches at least as much bed as face; they shall generally be at least four feet in length. The headers will be of similar size with the stretchers, and shall hold the size in the heart of the wall that they show on the face, and be so arranged as to occupy one fifth of the face of the wall, and they will be similarly disposed in the back. When the thickness of the wall necessitates it, stones not less than four feet in length will be placed transversely in the heart of the wall, to connect the two opposite sides of it. The stones for the heart of the wall need not be jointed, but must be well fitted to their places. The stones forming the points of piers which act as ice-breakers shall be neatly dressed on their faces; the other face-stones will, with the exception of the draft, be generally left with the face as they come from the quarry, unless the projections above the draft should exceed two inches, in which case they shall be roughly scabbled down to that point. The coping shall not be less than fourteen inches thick and three feet wide, well dressed and fastened together and to lower courses with clamps of iron, without extra charge.

54. Second-class Masonry.—This comes under squared-stone masonry (§§ 6 and 13), though the facing is generally a better grade of work. Extracts from specifications for this class are given below.

It shall be range-work or in broken courses, as may best suit the stone that is used. Face-stones to be accurately jointed and bedded, and no stone to be less than eight inches thick. The stretchers in the face to have beds of at least fifteen inches, and in no case less bed than rise, and to be not less than three feet long, measured in the face of the wall. The headers shall not have less than sixteen inches length of face, and shall extend at least three and one half feet into the wall, and not less than one header to every seven feet of wall, measured from centre to centre, and so arranged that a header in a superior course shall be placed between two headers in the course below. Backing-stones to be of large size, have parallel beds, but beds not dressed, as in first-class masonry. The outside stone to be set in cement mortar. Drafts to be cut on all corners of masonry.*

---

* For complete specifications on all grades of railroad masonry see "A Treatise on Masonry Construction," by Prof. I. O. Baker.

ART. 7.—CANAL-LOCK MASONRY.

PLATE XI.—ST. MARY'S FALLS CANAL-LOCK.

55. The drawings and specifications for this large and important piece of masonry are very detailed and very precise. Outside of government work such exhaustive drawings would hardly be furnished for similar construction.

Extracts from the specifications for the cut stone say:

" The cut stone shall conform approximately in dimensions to the bill of materials appended to these specifications, but exact drawings of the stones will be furnished the contractor when he needs them."

No joints were cut closer than $\frac{3}{8}$ inch, and it was specified that, " No joint shall vary in thickness from the dimensions specified, and the dimensions of all stones shall be such that the centre of each vertical joint shall not vary more than $\frac{1}{10}$ inch on either side of a vertical line through the centre of the lowest corresponding joint."

Drawings were therefore made of every course of stone in the lock (and there were twenty-three such courses), elevations, sections and patterns, and other details, specimens of which are shown on Plate XI. The full-size patterns for every stone not rectangular were cut out of zinc and furnished the contractor by the United States engineers in charge.

56. There being a large number of such patterns (471), some system of identification was of course necessary, and each different pattern received a distinctive number. The numbering on the plan and the patterns corresponding are

PLATE 11
800 FT. LOCK
ST. MARY'S FALLS CANAL, MICH.

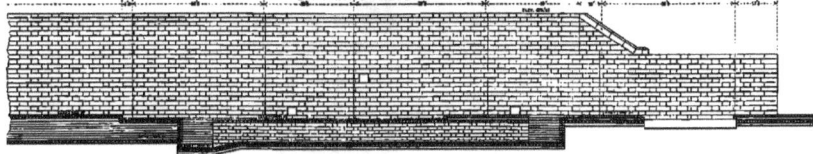

LONGITUDINAL SECTION OF FLOOR
AND
ELEVATION OF WALL AT EAST END

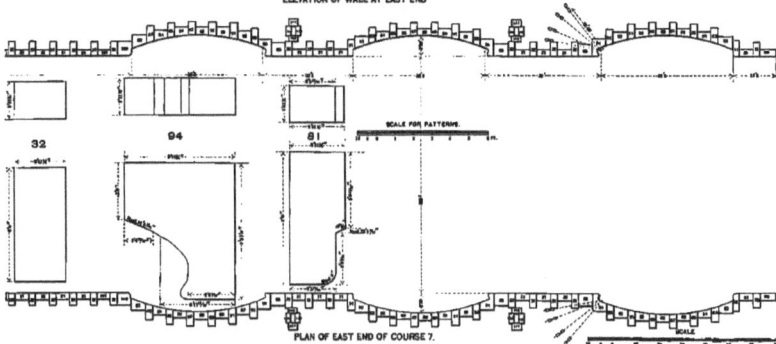

SCALE FOR PATTERNS.

32   94   81

PLAN OF EAST END OF COURSE 7.

SCALE

Back of
Foldout
Not Imaged

shown on this plate. An idea of the scheme followed for identifying the cut stone for a large building, as they are received from the stone-cutters, may be gained from Plates XII and XIII.

In the lock now under consideration the finished stone, when inspected by the United States inspector at the stone-yard, was marked with its proper pattern number, course number, and running number. The exact place on the wall for each stone of the successive courses as carried up was accurately marked for the contractor. The contractor, therefore, had only to cut the stone to certain patterns and put it on the place marked for it on the wall. There were over 1400 pieces of cut stone, aggregating nearly 19,000 cubic yards, and they all went together so nicely that it took less than ten hours' work for a stone-cutter to correct errors on the whole job.

57. The stones varied in size from 3½ ft. × 8 ft. × 8½ ft. to ⅞ ft. × 1⅛ ft. × 8 ft., giving 238 and 6 cu. ft. respectively. In general the dimensions were 2 ft. × 3 ft. × 6 ft. As the stone was specified to weigh not less than 148 lbs. to the cubic foot, the above dimensions give over 35,000 lbs. for the largest, and over 5000 lbs. for the average weight of each stone. These figures should give the student a good guide as to the dimensioning of cut stone, that they may be easily handled with the ordinary machinery.

It may be of interest to note that the cut-stone work in this structure was furnished and laid in the walls for $28.50 per cu. yd., a reasonable figure for such work.

ART. 8.—ARCHITECTURAL STONEWORK.

PLATE XII.—TOWER DOOR FROM THE BUILDING FOR LIBRARY OF CONGRESS.

58. The drawings on this and the next two plates are further examples of carefully detailed work. All the face-stones in this immense building are dimension-stones, and details like those on Plate XI were necessary. In this case, however, the usual practice was followed; namely, that of furnishing the contractor for the stonework with the dimensioned drawings instead of the patterns, and letting him make the latter himself. The inspected stone was properly marked for ready identification, shipped to the site of the building, and at once deposited near its final place.

59. Plate XII is a recessed entrance in a circular wall. The clear opening, being less than four feet, is readily spanned by a single stone and the construction of an arch as in Plate V avoided.

60. Exercises.—1. Draw a horizontal section through the course *G*.

2. Draw isometrics of the stones *d*, *b*, and *e*.

3. Draw the necessary patterns for *d*, *b*, and *e*.

4. Draw an isometric of, and the necessary patterns for, one of the stones in the doorway, in course *E*.

PLATE 12

TOWER DOOR

FROM THE

LIBRARY OF CONGRESS

SCALE

0 ⌐⌐⌐⌐ 1 Ft.

PLAN ON LINE T-T

5'-0" to Centre

1'-10"

BRICK WORK

ONE-STONE

BRICK WORK

COURSE H

COURSE G

BRICK WORK

BELT COURSE   COURSE H

COURSE G

COURSE F

COURSE E

COURSE D

COURSE C

COURSE B

COURSE A

BASE COURSE

BRICK

JOINT

JOINT

JOINT

JOINT

JOINT

JOINT

JOINT

JOINT

2"

1'-4"   1'-4"

2"   1'-4"   1'-4"   2"

1'-7½"

ELEVATION

R

LINTEL COURSE   1'-4"   2'-6"

3'-6"

SILL   9"

BRICK WORK

SECTION R-R

5'-0"

1'-10"

BASE COURSE

BASE COURSE

1'-4"   1'-4"

1'-6"   1'-6"

2'-3"   2'-3"

PLAN ON LINE S-S

PLATE 13
ENTRANCE & PORTE COCHÈRE
FROM THE
LIBRARY OF CONGRESS
SCALE OF FEET
0   1

ELEVATION

PLAN

NOTE: FIGURES FOR STEPS & PLATFORMS TAKEN TO FACES OF RISERS AND PLATFORMS, NOT TO NOSING

PLATES XIII AND XIV.—ENTRANCE AND PORTE COCHÈRE FROM BUILDING FOR LIBRARY OF CONGRESS.

61. These drawings show a complex and elegant piece of architectural masonry. It is impossible in this limited space to give all the necessary detail drawings, but a sufficient number are given to show the student what else would be required. Every joint of the stones is carefully shown, so that the full-size patterns can easily be made, and the dimensions extending into the walls are given by numerous sections. Some of those not shown are horizontal sections of the courses marked R, S, T, and U on the elevation, and vertical sections at different points along the railing, etc.

62. The development of the railing is obtained very simply, thus: From the point A on the plan as an origin the length of tread No. 1 as measured on the outer curve of the rail is laid off in the development; 2, 3, 4, etc., are similarly laid off, and the vertical distances of the railing above the plane of each step as obtained from the elevation marked on the corresponding places of the development. The pattern of each stone of the railing is now easily obtained by bending the developed pieces to the proper curves shown on the plan. The faces are of course left rough, so that the cutter who does the artistic work may draw and cut the ornaments.

63. Exercises.—1. Make details, isometric and patterns, for one of the railing-posts.

2. How would the stone-cutter work the skew part of the top of the railing?

PLATE 14
ENTRANCE & PORTE COCHERE
FROM THE
LIBRARY OF CONGRESS

SCALE OF FEET

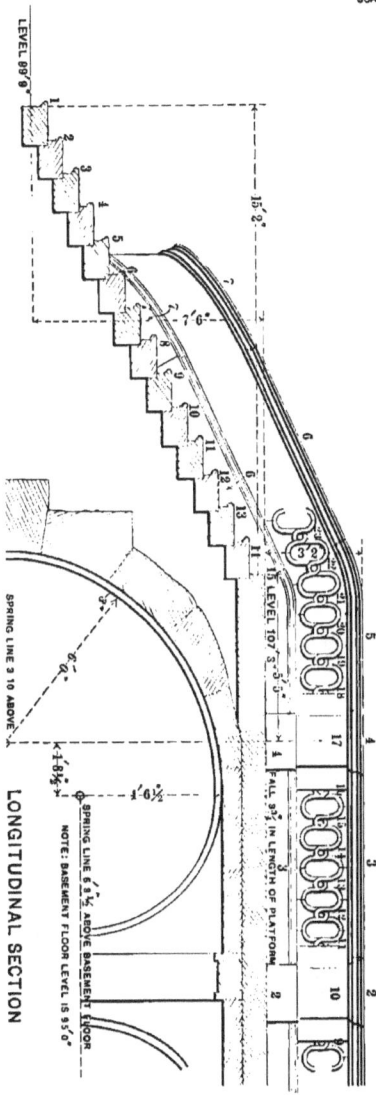

ELEVATION OF ENTRANCE STEPS

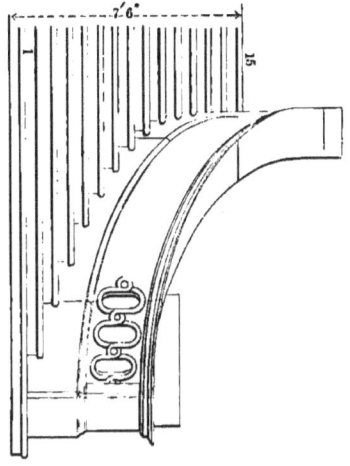

DEVELOPMENT OF RAILING

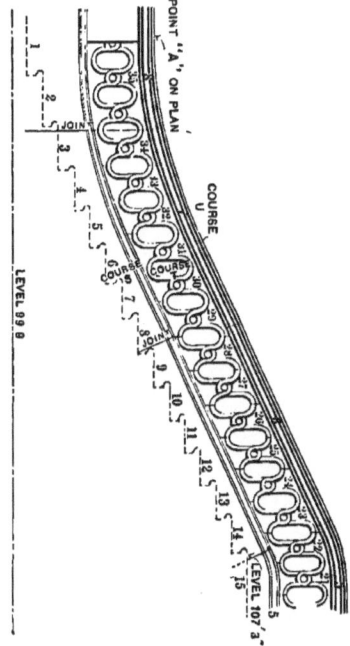

LONGITUDINAL SECTION